Made in United States
North Haven, CT
12 November 2026

22222923R00054